Aberdeen by Tram
in photographs by ENC Haywood

Bridge of Don terminus, 16th March 1958.

© Stenlake Publishing, ENC Haywood, 2024
First published in the United Kingdom, 2024,
by Stenlake Publishing Ltd.
www.stenlake.co.uk
ISBN 978-1-84033-969-7

The publishers regret that they cannot supply copies of any pictures featured in this book.

Printed by
P2D Books, 1 Newlands Rd, Westoning, Bedford, MK45 5LD

Union Street, 19th January 1958.

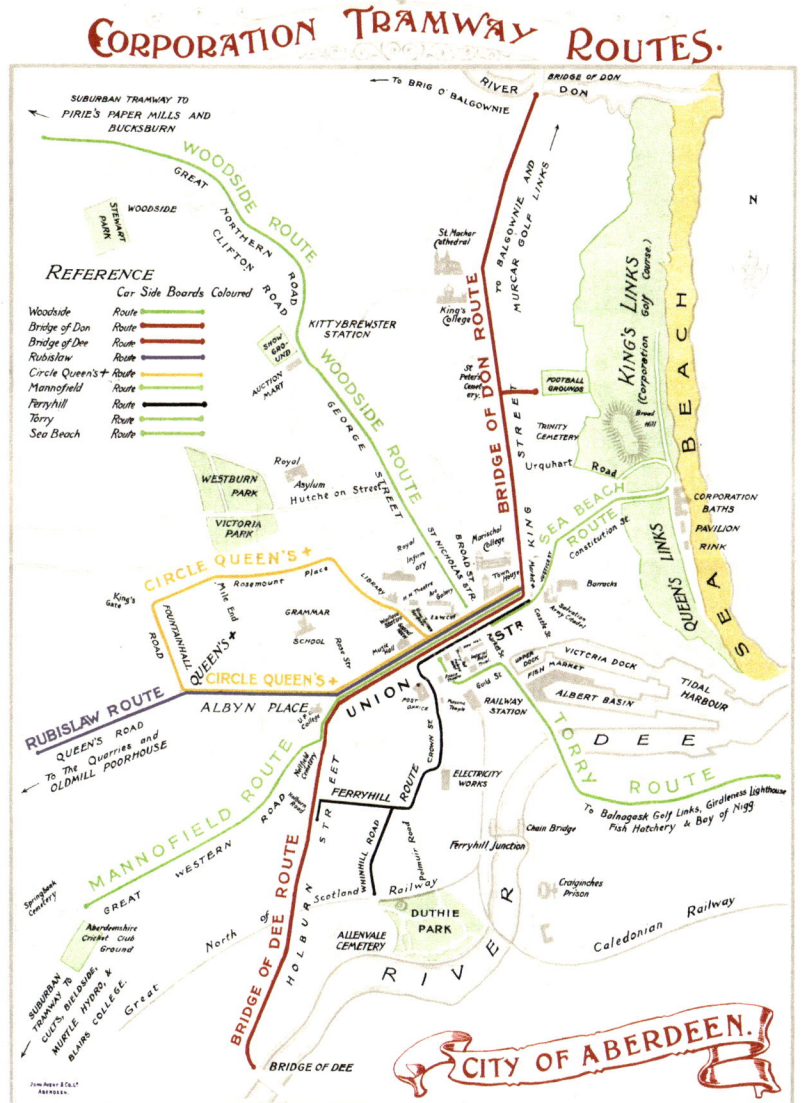

Introduction

A horse-drawn tramway in Aberdeen was established in 1874. By 1898 the network was in need of modernisation and was purchased by Aberdeen Corporation who upgraded and modernised the system. The map on the left, from before the First World War, shows the tramways at their greatest extent. There were also two lines operated by the Aberdeen Suburban Tramway company that linked Bucksburn to Woodside and Bieldside to Mannofield. These were the first part of Aberdeen's tram network to close in 1928 when the Corporation stopped running through services from the city, saving on the cost of modernisation but driving the trade to the competing bus companies. A combination of congestion and bus competition saw the demise, in 1931, of the tram routes to Torry (February), and Ferryhill (May). After the closures the tramway settled down, new more modern trams were introduced, and the network was divided into several routes.

1. From Bridge of Don to Bridge of Dee
2. Mannofield to Castle Street
3. Circle route via Queens Cross (clockwise)
4. Hazelhead to Castle Street
5. Kings Gate to Castle Street
6. Circle route via Queens Cross (anticlockwise)
7. Scatterburn at Woodside to St Nicholas Street
9. Castle Street to the Beach.

The Corporation ran a No. 8 bus, which was never a tram route.

This was the system when ENC Haywood began taking photographs of the city's trams. A native of Nottingham, he spent at least one holiday every year with relatives in Aberdeen. He loved the city, and he loved trams, taking the opportunity on his journeys between Nottingham and Aberdeen in the late 1940s, 50s and 60s to photograph the UK's surviving tram networks, as well as railways and other industry. He was perhaps spurred on by the fact that many of the Nottingham trams, whose network closed in 1936, had been purchased by Aberdeen Corporation. However, in Aberdeen too, the end of the tram network was being considered. 1951 saw the end of Route 2 to Mannofield, Routes 3, 6 and 5 in 1954, Route 7 in 1955, Route 4/9 in 1956, and finally the closure of the network in 1958.

Looking up Holburn Street from Bridge of Dee terminus, 7th February 1958.

Holburn Street at Ruthrieston Road, 4th April 1958

Bridge carrying Deeside line over Holburn Street looking toward city, April 1958.

Holburn Street at Ashvale Place, April 1958.

Alford Place, Holburn Street, Union Street junction, line to Queens Cross on right, April 1957.

Union Street at Union Bridge, 12th April 1957.

Castle Street and Market Cross. 4 34

Castle Street lokning north along King Street, April 1958.

Tramway depot, King Street (left), 19th January 1958.

Tram with Pittodrie Street behind, 3rd May 1957.

Trams in Merkland Road East waiting for end of football match, 1948.

At the Bridge of Don terminus, 20th February 1958.

At the Mannofield terminus, the old tramway depot behind the tram. Great Western Road at Duthie Road, June 1948.

At the Mannofield terminus from the opposite side of Great Western Road, looking to Countesswells Road, June 1948.

At Queen's Cross looking down Albyn Place. June 1948.

At Queen's Cross, June 1954.

Turning into Union Terrace from Union Street, June 1954.

Turning into Rosemount place from South Mount Street, June 1954.

Hazlehead track to Park, June 1948.

Queen's Road turn onto Hazlehead Park track, June 1954.

At Queen's Cross, June 1954.

Castle Street looking west. The tram is looping round the Mercat Cross before heading back to the beach. 20th August 1955.

Bridge of Don-bound tram. The track going straight past the Salvation Army Citadel along Justice Street, is heading to the beach, 20th August 1955.

Justice Street at its junction with East North Street, July 1954.

Park Street at the curve where the tram line turned into Constitution Street, July 1954.

Bridge carrying Constitution Street over Harbour Branch of railway, July 1954.

Constitution Street passing through Queen's Links, the Beach Bathing Station ahead, June 1948.

Looking back to Aberdeen along Constitution Street, July 1954. The original 1930s plan for Beach Boulevard (on the left) would have rerouted the Sea Beach tram along its length, and the tracks for this stretch into the distance. However, the Second World War stopped construction and when it resumed in 1956 trams were no longer part of the plan. By the time the Boulevard opened in May 1959 the tram network had closed.

At the Beach terminus, in the background is the Bathing Station now shorn of its northern wing, July 1954.

Terminus looking towards the end, June 1948.

Turning into Union Grove, June 1954.

Beechgrove Terrace passing Beechgrove Avenue, June 1954.

Beechgrove Terrace at King's Gate terminus, tram using crossover to head back into city, June 1954.

In Great Northern Road looking east from roughly where the Murray Court tower block is now, 22nd April 1955.

At the Fountain in Great Northern Road, 19th March 1955.

With the Fountain in background Great Northern Road tram depot, June 1954.

Astoria Cinema and Northern Hotel, Kittybrewster, June 1954.

Bridge over railway, Kittybrewster, June 1954.

Bridge over the railway at Kittybrewster, looking towards city, June 1954.

George Street Powis Place junction. June 1954.

At Loch Street where the track became single as it headed along George Street into the city, June 1948.

George Street becomes St Nicholas Street at its junction with Schoolhill, where there was a tram passing place, June 1948.

Second passing place where entrance to St Nicholas churches located, June 1948.

St. Nicholas terminus just bveyond the second passing place, 21st August 1955.

Leaving St Nicholas Street and heading along Union Street towards Castle Street, June 1954.